2020

_____

_____

"Prosperity is your birthright, and you hold the key to more abundance – in every area of your life – than you can possibly imagine."

2020

January

Su	Mo	Tu	We	Th	Fr	Sa
			1	2	3	4
5	6	7	8	9	10	11
12	13	14	15	16	17	18
19	**20**	21	22	23	24	25
26	27	28	29	30	31	

February

Su	Mo	Tu	We	Th	Fr	Sa
						1
2	3	4	5	6	7	8
9	10	11	12	13	**14**	15
16	**17**	18	19	20	21	22
23	24	25	26	27	28	29

March

Su	Mo	Tu	We	Th	Fr	Sa
1	2	3	4	5	6	7
8	9	10	11	12	13	14
15	16	17	18	19	20	21
22	23	24	25	26	27	28
29	30	31				

April

Su	Mo	Tu	We	Th	Fr	Sa
			1	2	3	4
5	6	7	8	9	**10**	11
12	13	14	15	16	17	18
19	20	21	22	23	24	25
26	27	28	29	30		

May

Su	Mo	Tu	We	Th	Fr	Sa
					1	2
3	4	5	6	7	8	9
10	11	12	13	14	15	16
17	18	19	20	21	22	23
24	**25**	26	27	28	29	30
31						

June

Su	Mo	Tu	We	Th	Fr	Sa
	1	2	3	4	5	6
7	8	9	10	11	12	13
14	15	16	17	18	19	20
21	22	23	24	25	26	27
28	29	30				

July

Su	Mo	Tu	We	Th	Fr	Sa
			1	2	**3**	**4**
5	6	7	8	9	10	11
12	13	14	15	16	17	18
19	20	21	22	23	24	25
26	27	28	29	30	31	

August

Su	Mo	Tu	We	Th	Fr	Sa
						1
2	3	4	5	6	7	8
9	10	11	12	13	14	15
16	17	18	19	20	21	22
23	24	25	26	27	28	29
30	31					

September

Su	Mo	Tu	We	Th	Fr	Sa
		1	2	3	4	5
6	**7**	8	9	10	11	12
13	14	15	16	17	18	19
20	21	22	23	24	25	26
27	28	29	30			

October

Su	Mo	Tu	We	Th	Fr	Sa
				1	2	3
4	5	6	7	8	9	10
11	**12**	13	14	15	16	17
18	19	20	21	22	23	24
25	26	27	28	29	30	**31**

November

Su	Mo	Tu	We	Th	Fr	Sa
1	2	3	4	5	6	7
8	9	10	**11**	12	13	14
15	16	17	18	19	20	21
22	23	24	25	**26**	27	28
29	30					

December

Su	Mo	Tu	We	Th	Fr	Sa
		1	2	3	4	5
6	7	8	9	10	11	12
13	14	15	16	17	18	19
20	21	22	23	24	**25**	26
27	28	29	30	31		

USA Holidays and Observances

Jan 01	New Year's Day	Jan 20	M L King Day	Feb 14	Valentine's Day
Feb 17	Presidents' Day	Apr 10	Good Friday	Apr 12	Easter Sunday
May 10	Mother's Day	May 25	Memorial Day	Jun 21	Father's Day
Jul 03	Independence Day Holiday	Jul 04	Independence Day	Sep 07	Labor Day
Oct 12	Columbus Day	Oct 31	Halloween	Nov 11	Veterans Day
Nov 26	Thanksgiving Day	Dec 25	Christmas		

2021

January

Su	Mo	Tu	We	Th	Fr	Sa
					1	2
3	4	5	6	7	8	9
10	11	12	13	14	15	16
17	**18**	19	20	21	22	23
24	25	26	27	28	29	30
31						

February

Su	Mo	Tu	We	Th	Fr	Sa
	1	2	3	4	5	6
7	8	9	10	11	12	13
14	**15**	16	17	18	19	20
21	22	23	24	25	26	27
28						

March

Su	Mo	Tu	We	Th	Fr	Sa
	1	2	3	4	5	6
7	8	9	10	11	12	13
14	15	16	17	18	19	20
21	22	23	24	25	26	27
28	29	30	31			

April

Su	Mo	Tu	We	Th	Fr	Sa
				1	**2**	3
4	5	6	7	8	9	10
11	12	13	14	15	16	17
18	19	20	21	22	23	24
25	26	27	28	29	30	

May

Su	Mo	Tu	We	Th	Fr	Sa
						1
2	3	4	5	6	7	8
9	10	11	12	13	14	15
16	17	18	19	20	21	22
23	24	25	26	27	28	29
30	**31**					

June

Su	Mo	Tu	We	Th	Fr	Sa
		1	2	3	**4**	5
6	7	8	9	10	11	12
13	14	15	16	17	18	19
20	21	22	23	24	25	26
27	28	29	30			

July

Su	Mo	Tu	We	Th	Fr	Sa
				1	2	3
4	**5**	6	7	8	9	10
11	12	13	14	15	16	17
18	19	20	21	22	23	24
25	26	27	28	29	30	31

August

Su	Mo	Tu	We	Th	Fr	Sa
1	2	3	4	5	6	7
8	9	10	11	12	13	14
15	16	17	18	19	20	21
22	23	24	25	26	27	28
29	30	31				

September

Su	Mo	Tu	We	Th	Fr	Sa
			1	2	3	4
5	**6**	7	8	9	10	11
12	13	14	15	16	17	18
19	20	21	22	23	24	25
26	27	28	29	30		

October

Su	Mo	Tu	We	Th	Fr	Sa
					1	2
3	4	5	6	7	8	9
10	**11**	12	13	14	15	16
17	18	19	20	21	22	23
24	25	26	27	28	29	30
31						

November

Su	Mo	Tu	We	Th	Fr	Sa
	1	2	3	4	5	6
7	8	9	10	**11**	12	13
14	15	16	17	18	19	20
21	22	23	24	**25**	26	27
28	29	30				

December

Su	Mo	Tu	We	Th	Fr	Sa
			1	2	3	4
5	6	7	8	9	10	11
12	13	14	15	16	17	18
19	20	21	22	23	24	**25**
26	27	28	29	30	31	

USA Holidays and Observances

Jan 01	New Year's Day	Jan 18	M L King Day	Feb 14	Valentine's Day
Feb 15	Presidents' Day	Apr 02	Good Friday	Apr 04	Easter Sunday
May 09	Mother's Day	May 31	Memorial Day	Jun 04	National Donut Day
Jun 20	Father's Day	Jul 04	Independence Day	Jul 05	Independence Day Holiday
Sep 06	Labor Day	Oct 11	Columbus Day	Oct 31	Halloween
Nov 11	Veterans Day	Nov 25	Thanksgiving Day	Dec 25	Christmas

What This Planner Can Do For You

Do you want to have more money flowing into your life?
Are you working very hard but you are still broke?
Are you a business owner struggling to make more money from your business?

You could be working very hard, you could be trying all kinds of money-making methods, or practising the Law of Attraction for a while but you just don't see yourself getting richer. If you wonder what is wrong, the answer is that you have not tuned yourself to the vibration of money in order to receive more money.

This Money Attraction Planner is what you need to let money flow towards you. This is not something to be read and put aside. This is your Money Attraction Planner or workbook to allow you to turn knowledge into action and experience the results. This is a money manifestation action plan that is designed to turn you into a money magnet. It can be used for the whole year.

There is a difference between knowing and doing. Most books can teach you how to do something, but unless you are able to do what is taught on a consistent basis, and make it a habit, you will not see results.

This Money Attraction Planner is designed as a weekly planner for you to use for the whole year - Jan to Dec 2020. It is designed to help you to internalise the concepts and turn them into positive habits to help you become a money magnet. It helps you to raise your money consciousness so that you can begin attracting more money and abundance into your life. If you want to have more money in the year 2020, this is the planner for you. There are three parts in this planner.

The first part of the planner teaches you the concepts of money attraction. The second part contains dated weekly calendar sheets and the third part contains money attraction worksheets:
The weekly calendar sheets are as follows:
- 12-month calendar
 - 12 sheets of monthly calendar sheet
- 128 pages of Weekly calendar sheets side-by-side for easy writing and viewing (Mon to Wed and Thurs to Sun for Jan to Dec 2020)
- 12 sheets of Monthly Money Attraction Tracker Worksheets for you to practise and track the actions needed taught in the first part.
- PLUS pages to record your money goals, Money WOWs and income sources

Part 1: Three Components Knowledge of Money Attraction

The three components are:
1) Understanding that money is a form of energy
2) Removing money blocks to enhance the energy of money to flow to you
3) Techniques for maintaining the right vibration from your mind and body to align you to the vibration of the energy of money.

Part 2: Money Attraction Planner Tracker Pages And Weekly Calendar Sheets

For every end of the week, turn to the Money Attraction Tracker page for that month and check the boxes provided (see below). This page is designed to track your progress for developing money consciousness and to keep track of your money attraction habits that will increase your positive vibration to attract the energy of money. Full explanation on the usage and purpose can be found in Part I.

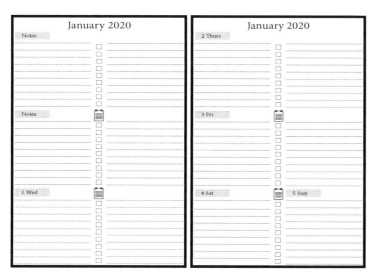

Money Attraction Tracker

128 pages of Weekly calendar sheets side-by-side for easy writing and viewing (Mon to Wed and Thurs to Sun for Jan to Dec 2020)

Money Attraction Tracker - Jan 2020

		Wk 1	Wk 2	Wk 3	Wk 4
Recite Money Goal		☐	☐	☐	☐
Recite Affirmations		☐	☐	☐	☐
Recite Mantra / Prayer		☐	☐	☐	☐
Practise Visualization	👁	☐	☐	☐	☐
Positive Outcomes		☐	☐	☐	☐
Emotion Meter Rating	☺	☐	☐	☐	☐
Gratitude Expressed	🎁	☐	☐	☐	☐
Money WOW	💲	☐	☐	☐	☐
Use Positive Words		☐	☐	☐	☐

Notes

Week 1	
Week 2	
Week 3	
Week 4	

Part 3: Essential Worksheets To Track Desires, Assets And Money Allocation

- My Desires List
- My Abundance Reality
- My Positive Mindsets
- My Money Affirmations
- My Money Sources
- My Assets And Investments
- My Money Allocation
- My Positive Words Bank

Let us begin.

Part 1: Money Is Energy

In order to have more money, you need to understand that money is energy.

Most of us see money as a physical form. We see money as pay checks, pieces of printed paper or metals. We equate money to salary or physical rewards for our hard work. But behind this physical form is energy.

To start having more, you need to perceive money as an expression of energy. When something is an energy, it is invisible, it flows, it repels, it attracts, it can be blocked and it can be unblocked.

What you cannot see, does not mean it does not exist. An electric current is invisible. If you touched a live wire, you will be electrocuted. Gravity is invisible. If Isaac Newton did not discover the Theory of Universal Gravitation, the force of gravity is still there whether if people really believe it or not. What you cannot see, does not mean it is not there.

In the field of quantum physics, everything around us is made up of energy. Our thoughts and desires are energy which can manifest itself as realities depending on its strength. The wooden chair you sit on may seem as a solid piece of wood in physical form, but if you can see deeper, you see it is a vibration of atoms vibrating at a certain frequency. What seems as solid to the human eye is alive with vibrations and energy. Our body itself is a mass of energy with billions of cells in our body vibrating all the time. Our mind is a massive energy source. So, everything is about energy and vibrations. This also means that the physical reality is a manifestation of your mental reality (energy and vibration).

If money is energy, then how can we have more of such energy? How can you make the money of energy flow towards you?

Money Blockages

First, you need to remove blockages that prevent you from receiving the energy of money. Blockages refer to:

- Negative money mindsets or beliefs from your subconscious mind
- Negative thoughts from your conscious mind
- Negative emotions such as guilt, hate, fear, jealousy, stress and anxiety
- Negative habits formed in the past

What are Negative Mindsets About Money

Negative beliefs about money in the subconscious mind are the most common blocks to the flow of money. There is a difference between knowing what you want and believing that you can have it. If you say you want more money, but in your mind you are thinking that it is not possible to get it, then it is not aligned. When your thoughts and your desires are not aligned, the universe will not give you what you want. The energy of money will not flow to you because of the misalignment.

Or when you say you want more money, but you hold the belief that money is the root of all evil, or rich people are bad people, or when you think that when you get too rich, you may mismanage it, then this is again not aligned.

To be able to let the energy of money flow to you, what you want and what your subconscious mind thinks must be congruent or aligned. This misalignment creates money blockage. So, once you cast away negative beliefs or mindsets about money, money will come to you.

The exercises recommended in this planner are to reprogram your mind, to reprogram your conscious and unconscious mind to make sure everything is aligned. It takes practice and time to change your beliefs and to cultivate good habits and thinking patterns. But first, you must let go of the old and welcome the new beliefs about money in your mind.

How to Overcome Negative Beliefs?

a) Replace With New Positive Beliefs

Embrace new beliefs and repeat them as often as you can to replace old beliefs.

Old Negative Beliefs	New Positive Beliefs
Money is hard to get	Money is abundant and there is enough for everyone

Rich people are bad people	Rich people are kind and charitable people
I can't handle my money	I am good at managing my money and I know how to grow it
Too much money is bad	Money can help the world a better place
I have to work very hard for money	Money comes to me easily

If you want to have more money, have more positive money beliefs.

b) Use Positive Affirmations

Affirmation is one way to reinforce new beliefs. Use positive affirmations about the nature of money and how you can receive the energy of money. Some affirmation statements are:

- I am a prosperity and money magnet
- Money flows to me easily from all directions
- I am becoming richer everyday
- More and more money is coming into my bank accounts
- I deserve to be rich
- I respect money and I think positive about money
- I can manage a lot of money and grow my wealth
- I believe that I have the talents and abilities to create unlimited wealth

A list of money affirmations is provided in this planner. Choose the ones that suit you and repeat them once in the morning and once before you sleep. You can use different affirmations for each week but make sure you repeat them once when you wake up and before you sleep. When you say it, make sure you attach a feeling to it. Say it with a positive emotion and it will be even more effective.

You may have been told that affirmation does not work. Yes, sometimes it may not work. This happens when your affirmations are phrased incorrectly. If your affirmation statement is not phrased correctly it becomes self-sabotaging and causes you to be out of alignment. For example:

I have a million dollars in my bank account.
Now, this is not a good statement because it is not believable by both your conscious and unconscious mind. When you look into your bank account, it does not show a million dollars. But if you say something like:

My money in my bank account is growing every day.
The second statement is more believable.

Or you say:
I believe in my talents and abilities to bring in more and more money into my life.

So, take note of how you phrase your affirmation. Your subconscious mind and conscious mind must agree to the suggestion from the statement.

Replace Negative Emotions With Positive Emotions

We are not a saint nor a Buddha. It is inevitable that we will have negative emotions on a daily basis. Fear, worry, anger, anxiety, disappointment, sorrow, sadness, rage, grief and frustration, and even impatience are forms of negative emotions. In order to attract the energy of money, our emotions must be positive on a daily basis.

We are emotional beings and there will be times where we will encounter failures, unhappy events, and challenges in our lives that create negative emotions. Other than external factors, the brain itself can be a troublesome source of negativity.

The aim of the conscious brain is to help you function and it is always alert. It is also very protective and it wants to keep you away from danger. It can become so protective till at times it creates more worries, fears and negative scenarios for you even if you did not ask for it. It can add a lot more pessimism into your life other than those negative imprints and negative beliefs that are stored in your subconscious mind.

Now, this negativity is bad and a key blockage to the flow of money energy. To allow money to flow into your life, you have to radiate positive vibes. You have to cultivate positive emotions at all times.

To maintain positive emotion on a daily basis, you must learn the art of switching negative charged energy into positive energy as soon as it arises and not to let it dwell too long on your mind. So, how can we do this?

Being Aware

First, you need to practice awareness. When you are aware that you have a negative thought about something, you must either tell your brain to stop or erase it. You have to literally talk to your brain and say, "Thanks for your thought, but I want to erase it". One way is to use the words – Cancel and Delete. When a negative thought arise, say to yourself "**Cancel and delete**".

Do this often and you can program your mind to shift to something else or to focus on other more pleasant thoughts.

If you sense a negative thought such as anger, you need to switch quickly. You can think of something funny and laugh it off. Chant a mantra to stay calm and stay focused on the chanting, or to do something that can take your mind off the negativity. For some people, running in the park or doing a work out can help to direct the focus of the negative event and feeling to something else. It must be released as soon as it arises and not to let your mind keeps it for too long.

Once your own emotional state is negative, you are again misaligned with the energy of money. In the planner pages section, you are to rate your positive emotion on a daily basis. Ask "How Can I feel Even Better", if your day is not a very positive one. This makes you think of ways to change it the next time, if your day is filled with negative emotions. The quickest way to make it better is to deepen your reflection on things that you are grateful for.

The Power Of Gratitude

Being grateful is a powerful tool to cultivate positive emotions. Hence, you are reminded to reflect as many as you can the things that you are grateful for each day. This is to let you cultivate the habit of using this powerful tool to discharge your negative emotions.

Why having gratitude is so powerful?

First, it helps us to look at the goodness around us. It heightens our consciousness for seeing good and when we think more of the good, more of the good will come our way. Good attracts good.

Secondly, it cultivates the habit of being thankful. The universe will not give to people who are not appreciative. If you are thankful, the universe will give you even more, because it knows you can handle it. So, the more you are thankful of what you got, the more you will receive. So, be grateful for all that you got, open your eyes and hearts and give thanks to the people and things around you.

Forgiveness And Deep Seated Negative Emotions

There may be other dark sides hidden inside you in the form of guilt and hatred from the past. These are even more negative than the anger and frustrations caused by external factors in your daily lives. People who suffer from depression, chronic health issues, people who get angry easily, or get stressed easily are also people with deeper negativity acquired from the past. You need to acknowledge this and forgive yourself or others who are part of this guilt or hate.

If you never free yourself from guilt, blame, shame and hatred, these deep-seated evils are the ultimate blockages to receiving the energy of money. No amount of positive affirmations and no amount of gratitude shown can boost your positive vibration to the universe. How can one release these evils? The only way is to tell yourself to let go. You must now forgive yourself and forgive others.

Guilt Removal

I will take some inspiration from the late Louise Hay. She is known as one of the founders of the self-help movement. She has given lectures and written many books on healing of the body and mind and she has sold more than 50 million books and CDs. I extracted and tweaked some of her affirmations slightly on what to say to yourself when you are trying to remove guilt. The first thing to do is to acknowledge the guilt and next, is to forgive yourself or other persons involved and let the guilt go. Here is a short version which I find useful.

Say this to yourself.

I am ready to release all forms of guilt that is holding inside me.

I now release all my conscious or unconscious thoughts and feelings of guilt.

I release all the guilt inside my mind, body and energy field caused by my own wrongdoings, shocks, betrayals, traumas, and by known and unknown causes and free my mind, body and energy fields from all forms of guilt from today.

I allow the energy of guilt to flow out of my body now!

I forgive myself! I forgive others!

I now allow the energy of peace to flow through my mind, body and energy field!

My mind, body and energy field are peaceful and relaxed

I am peace, I am love.

All is well.

(Source:https://www.healing-affirmations.com/tag/affirmations-for-guilt/)

Removing Blame and Hate Towards Someone

If you have past hatred and blame on someone, release these evils now. Here is a simple method. Write down the name of the person on a piece of paper.
Say a prayer such as:

I now clear and release all my hatred and anger from my mind, body and energy fields that

I am holding on to this person.

I forgive this person and it is time to remove all anger and hatred on this person.

I allow the energy of anger and hatred to flow out of my body now!

It is time to allow my anger to cool and dissipate.

I now allow the energy of peace to flow through my mind, body and energy field!

I choose to be calm and serene!

I choose to free myself from all thoughts of anger or bitterness!

I am peaceful, I am loving, I am relaxed, I am comforted and I am reassured.

I am liberated and all is well.

Source: https://www.healing-affirmations.com/release-transform-anger/
(Note: Edited for length and context)

After you finished saying these words, burn the paper. By doing so, you transfer the bad energy to the paper instead of holding on. It may sound weird like a ritual but, it works great psychologically. Rituals are great for programming your subconscious mind.

What you broadcast, you will receive. What you send out, you will get.

The Power Of Visualization

Visualization is a powerful way to raise the energy level of your thoughts. Visualization activates your brain and body to work. It convinces your subconscious mind of the reality and it boosts your confidence and the power of that belief. The more vivid and real is the visualization, the more you are able to manifest that thought into reality. When you visualize, create images in your mind, add sounds, feelings and emotions into it to make it as real as possible. When you attach an emotion to it, visualization will be even more powerful. For example, when you visualise yourself receiving a $100,000 cheque, feel a sense of excitement and joy inside you and visualise how the cheque looks like (what color is it and the words written on it). You can even act it out by jumping with joy as you visualise it.

Try to practise daily visualization or a few times a week. You will visualise your money desire everyday if you can. Be as vivid as possible. A check box for each week is created for this visualization exercise. Make sure you check it during your weekly review. The most effective way to do it, is before you go to sleep and when you wake up in the morning. If not, it can be done anywhere (sitting in the train, on the bus, in a cab, drinking a cut of coffee at a café). If you have a smartphone, you can set an alarm reminder to remind you to do visualization everyday.

Visualisation works for any goals you set. Every beginning of a new year, many people write down their goals they want to achieve. But the majority will not realise their goals. The main reason is because most people write it down and never look at it again. They assume that they are doing what is needed to meet that goal on a daily basis but unless you remind and visualise your goal everyday, you will lose focus along the way. There is too much distractions in our daily life that will take us off the track to achieve our goals.

Daily Positive Outcomes Visualization

Do you remember more vividly the negative words that people have said to you or praises that you receive? Do you remember the good things that others have done for you or the bad things that they have done? The answer is more likely that we remember more of the bad than the good.

Why is this so? The answer is this thing called the brain negative bias. There is much research done on this aspect so I will not dwell into the evidences but you can Google the words brain negative bias to learn more. The cause of this negative bias from our brain is to protect us from being harmed. It retains more negative imprints than positive ones. It reacts more strongly to negative stimuli than positive ones. Your brain is wired to be alert and always on the lookout for danger. But with such a mechanism it also means that it can cause one to be more negative in our perspective in the longer term. Now that you understand this, it is time to reduce this impact of this natural mechanism in your brain.

In this planner, I have created a section to let you visualise the good than to always think of the negative. On a weekly basis reflect the good things that have happen and check the box in the Weekly tracking sheet. Another way to practise this is to choose a task for each day and write out in one sentence, the positive outcome of the task before it actually gets done. For example, if you are meeting a new customer, imagine that the meeting went smoothly and it is a joy meeting him or her.

So, you write down:
I met Jason today, and he is such a pleasant guy, he is really keen to use our company product.

Do this on a daily basis and you will get into the habit of thinking positive outcomes than to imagine the worst each time. Remember, when you set the stage for success, you will manifest it. What you conceive, is what you get.

The Power of Words

Do you have a habit of using positive words or negative words? Words have power. What you say is what you get. Besides positive visualization, another powerful way to send out good vibration is to use more positive words. We know that a sound is a vibration; it has a frequency and a sound wave. Sound is energy. So, every time you utter a word, you are sending out a vibration and energy. Cultivate the habit of using positive words. Change the words when you speak and you can change your life. Stop complaining, stop gossiping, and stop useless and mindless chatters. Avoid the habit of retelling something bad that happened to you. The more you tell it, the more of the same thing you will attract. You may feel that you need to let off steam and anger but, by telling it over and over again, the more you will focus on it and the more negativity you will multiply. If there is nothing better to say, remember to stay silent and be at ease. Silence is definitely golden if your words are worth nothing good.

Make sure you refer to the list of powerful positive words in this planner and use as many as you can each day. To help you cultivate this habit, you will pick 3 positive words you want to use for each day and make sure you use them for that day. This exercise is to create an awareness of the words you use on a daily basis. The more positive words you use, the better your vibration.

Money Management

Money needs to flow and be respected to receive more.

When you start managing it consciously, you are activating the energy of money to flow to you. Money flows to those who can manage it and respect it. When you start managing your money and assets, you are raising your money consciousness. If you have dormant bank accounts, decide if you want to keep it or close it. Bank accounts should be active to allow the energy of money to flow freely.

Your first task is to write down all your money accounts in the My Assets And Investments page in this planner.

Write down the Bank Name, Bank Balance, for all your savings, current, retirement accounts and fixed deposit accounts. Write down all assets you currently own. Assets like investment, stocks and shares and properties.

List down all sources of income. Sources such as:
-       Monthly Salary
-       Part-time Salary

- Rental Income
- Income and dividends from investment
- Gift or inheritance
- Social Welfare Benefits
- Internet income sources (Amazon, Ebay, PayPal account or similar)
- Unexpected Income Source

List down and update it on a monthly basis to keep them active in your mind.

Make a habit to respect money of any kind and from any source. When you receive money, celebrate it, say THANK YOU. When you receive a check, say Thank You and keep it safe and deposit it immediately. Give priority to deposit the checks you received and not leave them lying around in the house. Show respect for money of any form and give it your full attention.

Keep all your bank books, check books, bank statements, share certificates and all money related documents in an organised and clean place. Get them organised and make sure you file these documents properly. These documents include monthly bills as well. Make sure all financial documents, bills, vouchers, insurance documents, and all kinds are filed neatly and properly. If you have not done this, do it now. Give it your priority.

Practise Receiving And Money WOW!

When you use this planner, you get to practise this good habit of celebrating money inflow and when receiving. There is a section called **Money WOW!**. This is where you record money or gifts that you receive. If you receive a check in your mailbox, write it down. If you receive money in your PayPal account when someone buys a product online, write it down. If you find money on the floor, pick it up and say Thank You!. Give thanks each time and say I LOVE MONEY, MONEY COMES TO ME EASILY! When you receive a gift from a friend or someone record it and say THANK YOU I LOVE RECEIVING GIFTS. This is to put you in the receiving mode so you can open up the channel to receive more. If someone wants to give you a treat, take it and thank the person and write it down too in this journal.

How many times can you recall when someone gives you a gift and you said things like "Oh, you shouldn't have.."

Do you think the giver is happy?

Of course not! When someone gives you a present, it is rude not to accept. Next time, just say thank you and receive it with joy. If someone wants to buy you a dinner or coffee, say YES!. Practise saying YES, to receive. Receive with grace and gratitude and you also allow the giver the chance to practise giving.

The point is not only must we practise asking, but we must also practise receiving. If you don't want to receive, then the universe will pass it over to the next person who wants it. The more you learn to receive, the more and better things will come to you.

Here are more tips for money management to let more energy of money flowing to you.

1. Keep proper accounting records for your business and your own cashflow records
2. File bills properly and pay them on time
3. Spend wisely and on what you need (hoarding or fear of spending are negative vibration)
4. Review your spending and cut unwanted spending
5. Pay off current and long term debts as fast as you can
6. Collect all money that people owe to you

Money Allocation Strategy

To enhance money flow, you need to cultivate the habit of allocating your money. From today, every sum of money you received either in cash, as a check, via bank deposit or in your PayPal account make a habit to allocate as follows:

- X% savings account
- X% for paying bills
- X% to give away to charity
- X% for My Happiness Account

What is **My Happiness Account**?
This is the account where you put aside money for you to spend on yourself to make yourself happy. If you like a shirt you saw at the mall, buy it. You can use it immediately after you allocated for it or let it accumulates every month and then spend it. The idea is not to keep it for too long because you want to experience the joy of spending money. Of course, I am not asking you to become a big spender and spend recklessly.

For example, if you receive a $100 check, and assuming your allocation percentage is as follows, then this is how much you set aside for each purpose:
- 40% savings account   = $40
- 40% for paying bills   = $40
- 10% to give away to charity   = $10
- 10% for My Happiness Account = $10

Money is energy, when you do any of these allocation, you are making the energy active and increasing the flow.

The Act Of Giving And Money Flow

The act of giving is a power way to shift from scarcity to an abundance mindset. By giving away some money, it means you have more than you need. The amount is not important. You can give 10 cents, 20 cents, a dollar, it does not matter. If you see a beggar in the street, give some money. But when you give, give it whole heartedly, with joy, loving-kindness and compassion, it sends out positive energy. The act of giving is aligned with the thought of abundance. Do not ask someone to give on your behalf. Give it personally and sincerely is the effective way.

Keep track of your acts of giving and see more money coming to you. Remember, money is energy, it cannot be trapped, there must be in-flow and out-flow.

Decluttering And Energy Boosting

Your physical environment is a reflection of what is inside you. If your living space is full of clutter and in a mess, very likely your life is a mess. Clutter blocks energy flow and this applies to the energy of money. If your working area is messy, clean it up now. Start NOW and I mean NOW!

If you have a messy living area, you are telling the universe that you are not ready to receive yet. Clean up. Make space for the money of energy to come in, clear up the mess and you will leave space for the universe to send you what you desire. Start clearing physical clutter today.

Besides removing clutter from your living space, remember to declutter your wallet. If your wallet is old and torn, get a new one. Do not stuff tiny bits and pieces of waste paper, name cards and receipts in your wallet. Keep it clean all the time and fill it with money. Keep your dollar notes in order by having notes arranged in order from small to big. Never leaves it empty. Once you remove clutter in your wallet, you are allowing more money to flow into your wallet. Declutter your wallet everyday from now on.

Weekly Review Exercise

For everything you do, you need to review your progress. Do your own weekly review.

Every end of the week, review the following:

1) Did you recite your affirmation consistently?
2) Did you do your visualization consistently?
3) Did you read out your desire?
5) Did you imagine positive outcomes for the things you are about to do?
6) How well did you maintain your positive emotions?
7) Did you practise gratitude?
8) Did you celebrate your Money Wow?
9) Did you watch your words and use more positive words and phrases on a daily basis?
10) Did you practise decluttering?
11) Did you allocate your money sources?
12) Did you manage your financial matters?
13) Did you give?
14) Did you practise receiving?
15) Did you spend your Happiness Account?
16) How can I do better?

What About Actions?

This planner is to help you focus on raising your money consciousness and vibration to attract money. It is designed to work on your inner being. The inner being shapes the outer being.

There is no need to rush into taking unnecessary actions to make you busy and tired. Do what you are doing now and put in your best effort. Do not rush into new ventures blindly. Maintain what you are good at doing and do them well. More money and opportunities will come by if you focus on sending out positive vibration. Look after your body, eat healthy and get enough rest and maintain your positive energy.

Should I buy lottery tickets? You can buy lottery tickets but buy the minimum. If you buy more, you are sending the signal that you don't trust yourself and the universe to let you win. Buying many tickets is a sign of scarcity mindset.

However, take note of sudden sparks of ideas that come to you just when you just woke up or when you are most relaxed. Pay attention to them and act on them. For example, you may have an idea that you need to go to a certain place then proceed with the idea. The universe could be sending you a good opportunity. If an image of someone, whom you have not met for a while, flashes across your mind, then you may want to contact the person. When you are vibrating positively, your intuition becomes more sensitive and you are able to receive messages from the universe. Whatever the universe wants you to do are often done with ease and with least resistance. Life is not a struggle when you are in the flow of positive vibration.

What is taught here will now become your new habits and routine. Just keep focusing on these new habits and sending out positive vibrations to attract the energy of money towards you.

Congratulations! You Are A Money Magnet!

Summary Of Actions To Take And How To Use The Review Page

Before You Start:

Before you start your money attraction exercise, identify some of your negative mindsets about money and start replacing them with positive mindsets about money. To help you acquire new mindsets, you will use positive money affirmations to program your subconscious mind.

If you are holding on to guilts and deep-seated hatred, it is time to let go now. Release these evils now before you start as taught in Part I.

Do up the worksheets in Part III.

Weekly Tracking:

For the end of each week, turn to the Money Attraction Tracker page. Reflect if you have performed the suggested actions in Part 1.

Money Attraction Tracker - Jan 2020	Wk 1	Wk 2	Wk 3	Wk 4
Recite Money Goal	☐	☐	☐	☐
Recite Affirmations	☐	☐	☐	☐
Recite Mantra / Prayer	☐	☐	☐	☐
Practise Visualization	☐	☐	☐	☐
Positive Outcomes	☐	☐	☐	☐
Emotion Meter Rating	☐	☐	☐	☐
Gratitude Expressed	☐	☐	☐	☐
Money WOW	☐	☐	☐	☐
Use Positive Words	☐	☐	☐	☐

Notes

Week 1	
Week 2	
Week 3	
Week 4	

1) Recite Your Money Goal

Set a money goal with a specific timeline. It can be your desire to receive $1000, $5000, $10,000 or $100,000. This is to get you into the habit of focusing on your money goal on a daily basis. No one can guarantee if you will receive the exact amount after a certain time period but by setting a goal, you are signaling to the universe your money desire thereby raising your vibration to attract the energy of money. By setting a desire, it puts you into the receiving mode. Recite your money goal daily. You can recite any number of times each day or recite once in the morning and before bedtime before reciting your affirmations.

2) Recite Your Affirmations
You can repeat the same affirmations or use different affirmations for different days. It is up to you. Refer to Part III for a list of money affirmations or you can write your own. You must recite your affirmations at least twice a day. Once in the morning when you wake up and before you go to bed preferred. Make sure you check the boxes provided. You can recite these affirmations any number of times in the day as well.

3) Recite Mantra / Prayer
If you have a religion, you can chant a mantra or say a daily prayer as this helps you to focus and increase your vibration. If you chant a mantra or say a prayer on a daily basis, check the box provided on the tracking sheet. I shall leave this as optional since some readers may not have a religion.

4) Practise Visualization
Visualise achieving your money goal at least twice a day. Once in the morning and once before bedtime after you have done reciting your money goal. Be as vivid as you can and remember to inject a feeling into it to make it powerful. Check on the boxe on the tracking page. The secret of effective visualisation lies in the feeling.

5) Positive Outcomes. On a daily basis, pick one task that you need to do and visualise their positive outcomes as explained in Part 1. Check if you do it for the week.

6) Rate your emotions for the week. Give a number from 1 to 10. (Low to High)
Be aware of your emotions daily. Is it negative or positive. If it is not so good, ask how can you feel better. Write down in the box the rating.

7) Check if you reflect on the things you are grateful for.

8) Money WOW! If you receive a gift, a favour, money for that week, put a tick.

9) Did you use positive words? Put a tick if you did for that week.

Write down how you feel and how you are progressing in the Notes section if you want. This section is optional.

Each time you receive an income, write it down in the My Income Sources worksheet and do the income allocation using My Income Allocation Worksheet.

May The Energy Of Money Be With You!

Part 3:

My Desires List

What do you really want? Write down the amount you want to have. Write down the things that you want money to get for you. The universe will bring you what you want.

_____

_____

_____

_____

_____

_____

_____

_____

_____

_____

_____

_____

_____

_____

_____

_____

My Abundance Reality

Imagine now you have received $5 million. Descibe in detail how you feel, how your life changed, what would you do with the money, who and how this money will benefit you and people around you. Be as vivid as you can., As you write inject feelings into it and really imagine that you already have it.

_____

_____

_____

_____

_____

_____

_____

_____

_____

_____

_____

_____

_____

_____

_____

_____

My Positive Money Mindset

List down a number of positive money mindsets that you will embrace from now on.

1. There is emough money everywhere in the world.

2. I deserve to be rich.

3. Money comes to me easily.

4. I am able to manage lots and lots of money in my bank account.

5. There is enough money and abundnace in this world for each and everyone.

6. Money is good and it can help many people to improve their lives.

My Money Affirmations

Here are some money affirmations. Choose any one you like to recite on a daily basis.

1. I am a money and prosperity magnet
2. The energy of money is always drawn to me
3. I am prosperous and my prosperity grows everyday
4. Money is a wonderful friend of mine
5. My bank account is always filled with money
6. I attract money naturally
7. A constant flow of money comes from known and unknown sources
8. I can be rich and wealthy
9. All the money I need is flowing to me now
10. Whatever I do attracts more money
11. Everyday I am attracting more and more money
12. It is my birth right to be wealthy and to receive abundance
13. I am tapped into the universe supply of money
14. Money is all around me
15. Money comes me in all kinds of way
16. I have unlimited multiple streams of income
17. I am the master of wealth
18. I manage money wisely
19. I always have more than enough money
20. I am wealthy, rich , healthy and wise
21. Successful money making ideas always come to me
22. My income is constantly increasing
23. I earn and receive more money every single day
24. Money is attracted to me and I am attracted to money
25. I have more money than I ever dream possible
26. I always have more money to spend and share
27. I attract money opportunities effortlessly
28. I love the energy of money
29. My vibration is always attuned to wealth
30. Money is all around me and I have access to it
31. I am a money master
32. I am a master of wealth

My Money Sources

List down all sources of income. Sources such as:

- Monthly Salary
- Part-time Salary
- Rental Income
- Income and dividends from investment
- Gift or inheritance
- Social Welfare Benefits
- Internet income sources (Amazon, Ebay, PayPal account or similar)
- Unexpected Income Source or Money WOW!

You can make your own spreadsheet and keep a record if you want to. Here is a sample how it can look like.

Month	Income Sources For 2020					
	Monthly Salary	Part-time Salary				
Jan						
Feb						
Mar						
Apr						
May						
June						
July						
Aug						
Sept						
Oct						
Nov						
Dec						

My Assets And Investments

Write down the Bank Name, Bank Balance, for all your savings, current, retirement accounts and fixed deposit accounts. Write down all assets you currently own. Assets like investment, stocks and shares and properties.

You can also use a spreadsheet to have a full picture of your finances. Here is a simple example.

Bank 1						
Bank 2						
Bank 3						
Property						

My Money Allocation

From today, every sum of money you received either in cash, as a check, via bank deposit or in your PayPal account make a habit to allocate as follows:

- X% savings account
- X% for paying bills
- X% to give away to charity
- X% for My Happiness Account

If you run out of space, you can use a spreadsheet for recording. Instead of doing on a daily basis, you can do it weekly or monthyl basis after you master the habit of doing so.

Date	Amount Received $	Savings $	Bills $	Charity $	My Happiness A/C $

Date	Amount Received $	Savings $	Bills $	Charity $	My Happiness A/C $

My Positive Words Bank

Use positive words on a daily basis. Here are a list of positive words for your reference.

Active	Full	Rejuvenate
Abundant	Glow	Relax
Alive	Grin	Renew
Animated	Healed	Restore
Beaming	Healing	Robust
Beautiful	Healthful	Serenity
Brave	Healthy	Shine
Best	Heart	Sparkling
Bloom	Hearty	Spontaneous
Complete	Instantaneous	Stillness
Core	Joyful	Strong
Courageous	Light	Sustain
Cultivate	Lively	Therapeutic
Cure	Luminous	Thrive
Dazzling	Metamorphosis	Transform
Divine	Miracle	Upbeat
Effortless	Motivate	Vibrant
Energized	Natural	Vivacious
Energy	Nourish	Well
Enthusiastic	Nourished	Whole
Fit	Nurture	Wholesome
Fantastic	Nutritious	Wondrous
Flourish	Perfect	
Freedom	Refresh	

Notes

Notes

Important Dates

Important Dates

Important Dates

Notes

January 2020

Sun	Mon	Tue	Wed	Thu	Fri	Sat
			1 New Year's Day	2	3	4
5	6	7	8	9	10	11
12	13	14	15	16	17	18
19	20 Martin Luther King Jr.	21	22	23	24	25
26	27	28	29	30	31	

January 2020

Notes

1 Wed

January 2020

2 Thurs

☐
☐
☐
☐
☐
☐
☐
☐
☐

3 Fri

☐
☐
☐
☐
☐
☐
☐
☐
☐

4 Sat

5 Sun

☐
☐
☐
☐
☐
☐
☐
☐
☐
☐

January 2020

6 Mon

7 Tues

8 Wed

January 2020

9 Thurs

10 Fri

11 Sat

12 Sun

January 2020

13 Mon

14 Tues

15 Wed

January 2020

16 Thurs

17 Fri

18 Sat

19 Sun

January 2020

20 Mon

21 Tues

22 Wed

January 2020

January 2020

27 Mon

28 Tues

29 Wed

January - February 2020

30 Thurs

31 Fri

1 Sat

2 Sun

Money Attraction Tracker - Jan 2020

	Wk 1	Wk 2	Wk 3	Wk 4
Recite Money Goal	☐	☐	☐	☐
Recite Affirmations	☐	☐	☐	☐
Recite Mantra / Prayer	☐	☐	☐	☐
Practise Visualization	☐	☐	☐	☐
Positive Outcomes	☐	☐	☐	☐
Emotion Meter Rating	☐	☐	☐	☐
Gratitude Expressed	☐	☐	☐	☐
Money WOW	☐	☐	☐	☐
Use Positive Words	☐	☐	☐	☐

Notes

Week 1	
Week 1	
Week 2	
Week 3	
Week 4	

February 2020

Sun	Mon	Tue	Wed	Thu	Fri	Sat
						1
2 Groundhog Day	3	4	5	6	7	8
9	10	11	12	13	14 Valentine's Day	15
16	17 Presidents Day	18	19	20	21 Int'l. Mother Language	22
23	24	25	26	27	28	29

February 2020

3 Mon

4 Tues

5 Wed

February 2020

February 2020

10 Mon

11 Tues

12 Wed

February 2020

February 2020

17 Mon

☐
☐
☐
☐
☐
☐
☐
☐
☐

18 Tues

☐
☐
☐
☐
☐
☐
☐
☐
☐

19 Wed

☐
☐
☐
☐
☐
☐
☐
☐
☐

February 2020

February 2020

February - March 2020

27 Thurs

28 Fri

29 Sat

1 Sun

Money Attraction Tracker - Feb 2020

	Wk 1	Wk 2	Wk 3	Wk 4
Recite Money Goal	☐	☐	☐	☐
Recite Affirmations	☐	☐	☐	☐
Recite Mantra / Prayer	☐	☐	☐	☐
Practise Visualization	☐	☐	☐	☐
Positive Outcomes	☐	☐	☐	☐
Emotion Meter Rating				
Gratitude Expressed	☐	☐	☐	☐
Money WOW	☐	☐	☐	☐
Use Positive Words	☐	☐	☐	☐

Notes

Week 1	
Week 2	
Week 3	
Week 4	

March 2020

Sun	Mon	Tue	Wed	Thu	Fri	Sat
1	2	3	4	5	6	7
8 Daylight Saving Begins	9	10	11	12	13	14
15	16	17 Saint Patrick's Day	18	19	20 Spring Begins (Northern Hemisphere)	21
22	23	24 World Tuberculosis Day	25	26	27	28
29	30	31				

March 2020

☐
☐
☐
☐
☐
☐
☐
☐
☐

📝

☐
☐
☐
☐
☐
☐
☐
☐
☐

📝

☐
☐
☐
☐
☐
☐
☐
☐

March 2020

5 Thurs

6 Fri

7 Sat

8 Sun

March 2020

9 Mon

10 Tues

11 Wed

March 2020

March 2020

March 2020

☐
☐
☐
☐
☐
☐
☐
☐
☐

☐
☐
☐
☐
☐
☐
☐
☐

☐
☐
☐
☐
☐
☐
☐
☐

March 2020

23 Mon

24 Tues

25 Wed

March 2020

27 Fri

28 Sat

29 Sun

Money Attraction Tracker - Mar 2020

	Wk 1	Wk 2	Wk 3	Wk 4
Recite Money Goal	☐	☐	☐	☐
Recite Affirmations	☐	☐	☐	☐
Recite Mantra / Prayer	☐	☐	☐	☐
Practise Visualization	☐	☐	☐	☐
Positive Outcomes	☐	☐	☐	☐
Emotion Meter Rating	☐	☐	☐	☐
Gratitude Expressed	☐	☐	☐	☐
Money WOW	☐	☐	☐	☐
Use Positive Words	☐	☐	☐	☐

Notes

Week 1	
Week 2	
Week 3	
Week 4	

April 2020

Sun	Mon	Tue	Wed	Thu	Fri	Sat
			1	2	3	4
5	6	7	8	9	10 Good Friday	11
12 Easter	13	14	15 Tax Day (Taxes Due)	16	17	18
19	20	21	22 Administrative Professionals	23	24 Arbor Day	25
26	27	28	29 Duke Ellington Day	30		

March - April 2020

30 Mon

☐
☐
☐
☐
☐
☐
☐
☐
☐

31 Tues

☐
☐
☐
☐
☐
☐
☐
☐

1 Wed

☐
☐
☐
☐
☐
☐
☐
☐
☐

April 2020

2 Thurs

3 Fri

4 Sat

5 Sun

April 2020

6 Mon

7 Tues

8 Wed

April 2020

9 Thurs

10 Fri

11 Sat

12 Sun

April 2020

13 Mon

14 Tues

15 Wed

April 2020

16 Thurs

17 Fri

18 Sat

19 Sun

April 2020

20 Mon

21 Tues

22 Wed

April 2020

23 Thurs

24 Fri

25 Sat

26 Sun

April 2020

27 Mon

☐
☐
☐
☐
☐
☐
☐
☐
☐

28 Tues

☐
☐
☐
☐
☐
☐
☐
☐
☐

29 Wed

☐
☐
☐
☐
☐
☐
☐
☐
☐

April - May 2020

30 Thurs

1 Fri

2 Sat

3 Sun

Money Attraction Tracker - Apr 2020

	Wk 1	Wk 2	Wk 3	Wk 4
Recite Money Goal	☐	☐	☐	☐
Recite Affirmations	☐	☐	☐	☐
Recite Mantra / Prayer	☐	☐	☐	☐
Practise Visualization	☐	☐	☐	☐
Positive Outcomes	☐	☐	☐	☐
Emotion Meter Rating				
Gratitude Expressed	☐	☐	☐	☐
Money WOW	☐	☐	☐	☐
Use Positive Words	☐	☐	☐	☐

Notes

Week 1	
Week 2	
Week 3	
Week 4	

May 2020

Sun	Mon	Tue	Wed	Thu	Fri	Sat
					1	2
3	4	5 Cinco De Mayo	6	7	8	9
10 Mother's Day	11	12	13	14	15	16 Armed Forces Day
17	18	19	20	21	22	23
24	25 Memorial Day	26	27	28	29	30
31						

May 2020

4 Mon

5 Tues

6 Wed

May 2020

7 Thurs

8 Fri

9 Sat

10 Sun

May 2020

May 2020

May 2020

18 Mon

19 Tues

20 Wed

May 2020

☐
☐
☐
☐
☐
☐
☐
☐
☐

22 Fri

☐
☐
☐
☐
☐
☐
☐
☐

23 Sat   24 Sun

☐
☐
☐
☐
☐
☐
☐
☐
☐

May 2020

25 Mon

26 Tues

27 Wed

May 2020

28 Thurs

29 Fri

30 Sat

31 Sun

Money Attraction Tracker - May 2020

	Wk 1	Wk 2	Wk 3	Wk 4
Recite Money Goal	☐	☐	☐	☐
Recite Affirmations	☐	☐	☐	☐
Recite Mantra / Prayer	☐	☐	☐	☐
Practise Visualization	☐	☐	☐	☐
Positive Outcomes	☐	☐	☐	☐
Emotion Meter Rating	☐	☐	☐	☐
Gratitude Expressed	☐	☐	☐	☐
Money WOW	☐	☐	☐	☐
Use Positive Words	☐	☐	☐	☐

Notes

Week 1	
Week 2	
Week 3	
Week 4	

June 2020

Sun	Mon	Tue	Wed	Thu	Fri	Sat
	1	2	3	4	5	6
7	8	9	10	11	12	13
14 Flag Day	15	16	17	18	19	20 Summer Solstice
21 Father's Day	22	23	24	25	26	27
28	29	30				

June 2020

1 Mon

2 Tues

3 Wed

June 2020

4 Thurs

5 Fri

6 Sat

7 Sun

June 2020

8 Mon

9 Tues

10 Wed

June 2020

11 Thurs

12 Fri

13 Sat

14 Sun

June 2020

15 Mon

16 Tues

17 Wed

June 2020

18 Thurs

19 Fri

20 Sat

21 Sun

June 2020

22 Mon

23 Tues

24 Wed

June 2020

25 Thurs

26 Fri

27 Sat

28 Sun

Money Attraction Tracker - June 2020

	Wk 1	Wk 2	Wk 3	Wk 4
Recite Money Goal	☐	☐	☐	☐
Recite Affirmations	☐	☐	☐	☐
Recite Mantra / Prayer	☐	☐	☐	☐
Practise Visualization	☐	☐	☐	☐
Positive Outcomes	☐	☐	☐	☐
Emotion Meter Rating	☐	☐	☐	☐
Gratitude Expressed	☐	☐	☐	☐
Money WOW	☐	☐	☐	☐
Use Positive Words	☐	☐	☐	☐

Notes

Week 1	
Week 2	
Week 3	
Week 4	

July 2020

Sun	Mon	Tue	Wed	Thu	Fri	Sat
			1 Canada Day	2	3	4 Indep. Day
5	6	7	8	9	10	11
12	13	14	15	16	17	18
19	20	21	22	23	24	25
26	27	28	29	30	31	

June - July 2020

29 Mon

_____
_____
_____
_____
_____
_____
_____
_____

☐
☐
☐
☐
☐
☐
☐
☐

_____
_____
_____
_____
_____

30 Tues

_____
_____
_____
_____
_____
_____
_____
_____

☐
☐
☐
☐
☐
☐
☐
☐

_____
_____
_____
_____
_____

1 Wed

_____
_____
_____
_____
_____
_____
_____
_____

☐
☐
☐
☐
☐
☐
☐
☐

_____
_____
_____
_____
_____

July 2020

2 Thurs

3 Fri

4 Sat

5 Sun

July 2020

6 Mon

7 Tues

8 Wed

July 2020

☐
☐
☐
☐
☐
☐
☐
☐
☐

☐
☐
☐
☐
☐
☐
☐
☐
☐

☐
☐
☐
☐
☐
☐
☐
☐
☐

July 2020

☐

☐

July 2020

16 Thurs

17 Fri

18 Sat

19 Sun

July 2020

20 Mon

21 Tues

22 Wed

July 2020

July 2020

27 Mon

☐
☐
☐
☐
☐
☐
☐
☐
☐

28 Tues

☐
☐
☐
☐
☐
☐
☐
☐

29 Wed

☐
☐
☐
☐
☐
☐
☐
☐
☐

July -August 2020

30 Thurs

31 Fri

1 Sat

2 Sun

Money Attraction Tracker - July 2020

	Wk 1	Wk 2	Wk 3	Wk 4
Recite Money Goal	☐	☐	☐	☐
Recite Affirmations	☐	☐	☐	☐
Recite Mantra / Prayer	☐	☐	☐	☐
Practise Visualization	☐	☐	☐	☐
Positive Outcomes	☐	☐	☐	☐
Emotion Meter Rating	☐	☐	☐	☐
Gratitude Expressed	☐	☐	☐	☐
Money WOW	☐	☐	☐	☐
Use Positive Words	☐	☐	☐	☐

Notes

Week 1	
Week 2	
Week 3	
Week 4	

August 2020

Sun	Mon	Tue	Wed	Thu	Fri	Sat
						1 Nat'l. Girlfriend Day
2	3	4	5	6	7	8
9	10	11	12	13	14	15
16	17	18	19	20	21	22
23	24	25	26	27	28	29
30	31					

August 2020

3 Mon

4 Tues

5 Wed

August 2020

6 Thurs

7 Fri

8 Sat

9 Sun

August 2020

10 Mon

11 Tues

12 Wed

August 2020

13 Thurs

14 Fri

15 Sat

16 Sun

August 2020

_____
_____
_____
_____
_____
_____
_____
_____
_____

☐
☐
☐
☐
☐
☐
☐
☐
☐

_____
_____
_____
_____
_____
_____
_____
_____
_____

☐
☐
☐
☐
☐
☐
☐
☐

_____
_____
_____
_____
_____
_____
_____
_____
_____

☐
☐
☐
☐
☐
☐
☐
☐

August 2020

20 Thurs

21 Fri

22 Sat

23 Sun

August 2020

24 Mon

25 Tues

26 Wed

August 2020

27 Thurs

28 Fri

29 Sat

30 Sun

Money Attraction Tracker - Aug 2020

	Wk 1	Wk 2	Wk 3	Wk 4
Recite Money Goal	☐	☐	☐	☐
Recite Affirmations	☐	☐	☐	☐
Recite Mantra / Prayer	☐	☐	☐	☐
Practise Visualization	☐	☐	☐	☐
Positive Outcomes	☐	☐	☐	☐
Emotion Meter Rating	☐	☐	☐	☐
Gratitude Expressed	☐	☐	☐	☐
Money WOW	☐	☐	☐	☐
Use Positive Words	☐	☐	☐	☐

Notes

Week 1	
Week 2	
Week 3	
Week 4	

September 2020

Sun	Mon	Tue	Wed	Thu	Fri	Sat
		1	2	3	4	5
6	7 Labor Day	8	9	10	11 Patriot Day	12
13	14	15	16	17	18	19
20	21	22 Fall begins	23	24	25	26
27	28	29	30			

August - September 2020

31 Mon

1 Tues

2 Wed

September 2020

3 Thurs

4 Fri

5 Sat

6 Sun

September 2020

7 Mon

_____

☐

8 Tues

☐

9 Wed

☐

September 2020

September 2020

14 Mon

15 Tues

16 Wed

September 2020

17 Thurs

18 Fri

19 Sat

20 Sun

September 2020

September 2020

September 2020

_____
_____
_____
_____
_____
_____
_____
_____

☐
☐
☐
☐
☐
☐
☐
☐
☐

_____
_____
_____
_____
_____
_____
_____
_____
_____

☐
☐
☐
☐
☐
☐
☐
☐

_____
_____
_____
_____
_____
_____
_____
_____

☐
☐
☐
☐
☐
☐
☐
☐
☐

October 2020

1 Thurs

2 Fri

3 Sat

4 Sun

Money Attraction Tracker - Sept 2020

	Wk 1	Wk 2	Wk 3	Wk 4
Recite Money Goal	☐	☐	☐	☐
Recite Affirmations	☐	☐	☐	☐
Recite Mantra / Prayer	☐	☐	☐	☐
Practise Visualization	☐	☐	☐	☐
Positive Outcomes	☐	☐	☐	☐
Emotion Meter Rating				
Gratitude Expressed	☐	☐	☐	☐
Money WOW	☐	☐	☐	☐
Use Positive Words	☐	☐	☐	☐

Notes

Week 1	
Week 1	
Week 2	
Week 3	
Week 4	

October 2020

Sun	Mon	Tue	Wed	Thu	Fri	Sat
				1	2	3
4	5	6	7	8	9	10
11 Come Out Day	12 Columbus Day	13	14	15	16 Boss's Day	17
18	19	20	21	22	23	24
25	26	27	28	29	30	31 Halloween

October 2020

5 Mon

6 Tues

7 Wed

October 2020

October 2020

12 Mon

13 Tues

14 Wed

October 2020

October 2020

19 Mon

☐
☐
☐
☐
☐
☐
☐
☐

20 Tues

☐
☐
☐
☐
☐
☐
☐
☐

21 Wed

☐
☐
☐
☐
☐
☐
☐
☐

October 2020

22 Thurs

23 Fri

24 Sat

25 Sun

October 2020

26 Mon

27 Tues

28 Wed

October - November 2020

Money Attraction Tracker - Oct 2020

	Wk 1	Wk 2	Wk 3	Wk 4
Recite Money Goal	☐	☐	☐	☐
Recite Affirmations	☐	☐	☐	☐
Recite Mantra / Prayer	☐	☐	☐	☐
Practise Visualization	☐	☐	☐	☐
Positive Outcomes	☐	☐	☐	☐
Emotion Meter Rating	☐	☐	☐	☐
Gratitude Expressed	☐	☐	☐	☐
Money WOW	☐	☐	☐	☐
Use Positive Words	☐	☐	☐	☐

Notes

Week 1	
Week 2	
Week 3	
Week 4	

November 2020

Sun	Mon	Tue	Wed	Thu	Fri	Sat
1 Daylight Saving Time Ends	**2**	**3** Election Day	**4**	**5**	**6**	**7**
8	**9**	**10**	**11** Veterans Day	**12**	**13**	**14**
15	**16**	**17**	**18**	**19**	**20**	**21**
22	**23**	**24**	**25**	**26** Thanksgiving Day	**27**	**28**
29	**30**					

November 2020

2 Mon

☐
☐
☐
☐
☐
☐
☐
☐
☐

3 Tues

☐
☐
☐
☐
☐
☐
☐
☐
☐

4 Wed

☐
☐
☐
☐
☐
☐
☐
☐

November 2020

November 2020

9 Mon

10 Tues

11 Wed

November 2020

12 Thurs

_____

_____

_____

_____

_____

_____

_____

_____

_____

13 Fri

_____

_____

_____

_____

_____

_____

_____

_____

14 Sat

_____

_____

_____

_____

_____

_____

_____

_____

15 Sun

November 2020

16 Mon

17 Tues

18 Wed

November 2020

☐
☐
☐
☐
☐
☐
☐
☐
☐

☐
☐
☐
☐
☐
☐
☐
☐
☐

☐
☐
☐
☐
☐
☐
☐
☐
☐

November 2020

23 Mon

24 Tues

25 Wed

November 2020

Money Attraction Tracker - Nov 2020

	Wk 1	Wk 2	Wk 3	Wk 4
Recite Money Goal	☐	☐	☐	☐
Recite Affirmations	☐	☐	☐	☐
Recite Mantra / Prayer	☐	☐	☐	☐
Practise Visualization	☐	☐	☐	☐
Positive Outcomes	☐	☐	☐	☐
Emotion Meter Rating	☐	☐	☐	☐
Gratitude Expressed	☐	☐	☐	☐
Money WOW	☐	☐	☐	☐
Use Positive Words	☐	☐	☐	☐

Notes

Week 1	
Week 2	
Week 3	
Week 4	

December 2020

Sun	Mon	Tue	Wed	Thu	Fri	Sat
		1	2	3	4	5
6	7	8	9	10	11	12
13	14	15	16	17	18	19
20	21 Winter Solstice	22	23	24	25 Christmas	26
27	28	29	30	31		

November - December 2020

30 Mon

1 Tues

2 Wed

December 2020

3 Thurs

4 Fri

5 Sat

6 Sun

December 2020

7 Mon

8 Tues

9 Wed

December 2020

10 Thurs

☐
☐
☐
☐
☐
☐
☐
☐
☐

11 Fri

☐
☐
☐
☐
☐
☐
☐
☐
☐

12 Sat

13 Sun

☐
☐
☐
☐
☐
☐
☐
☐
☐

December 2020

14 Mon

☐
☐
☐
☐
☐
☐
☐
☐
☐

15 Tues

☐
☐
☐
☐
☐
☐
☐
☐
☐

16 Wed

☐
☐
☐
☐
☐
☐
☐
☐

December 2020

December 2020

21 Mon

22 Tues

23 Wed

December 2020

24 Thurs

25 Fri

26 Sat

27 Sun

December 2020

28 Mon

29 Tues

30 Wed

December - January 2021

31 Thurs

☐
☐
☐
☐
☐
☐
☐
☐
☐

1 Fri

☐
☐
☐
☐
☐
☐
☐
☐
☐

2 Sat   **3 Sun**

☐
☐
☐
☐
☐
☐
☐
☐
☐

Money Attraction Tracker - Dec 2020

	Wk 1	Wk 2	Wk 3	Wk 4
Recite Money Goal	☐	☐	☐	☐
Recite Affirmations	☐	☐	☐	☐
Recite Mantra / Prayer	☐	☐	☐	☐
Practise Visualization	☐	☐	☐	☐
Positive Outcomes	☐	☐	☐	☐
Emotion Meter Rating	☐	☐	☐	☐
Gratitude Expressed	☐	☐	☐	☐
Money WOW	☐	☐	☐	☐
Use Positive Words	☐	☐	☐	☐

Notes

Week 1	
Week 2	
Week 3	
Week 4	

Notes

Expense Tracker - Jan 2020

Date	Description	Amount	Payment Mode	Paid
				☐
				☐
				☐
				☐
				☐
				☐
				☐
				☐
				☐
				☐
				☐
				☐
				☐
				☐
				☐
				☐
				☐
				☐
				☐
				☐
				☐
				☐
Total Amount For This Month				

Expense Tracker - Feb 2020

Date	Description	Amount	Payment Mode	Paid
				☐
				☐
				☐
				☐
				☐
				☐
				☐
				☐
				☐
				☐
				☐
				☐
				☐
				☐
				☐
				☐
				☐
				☐
				☐
				☐
				☐
				☐
Total Amount For This Month				

Expense Tracker - Mar 2020

Date	Description	Amount	Payment Mode	Paid
				☐
				☐
				☐
				☐
				☐
				☐
				☐
				☐
				☐
				☐
				☐
				☐
				☐
				☐
				☐
				☐
				☐
				☐
				☐
				☐
				☐
Total Amount For This Month				

Expense Tracker - Apr 2020

Date	Description	Amount	Payment Mode	Paid
				☐
				☐
				☐
				☐
				☐
				☐
				☐
				☐
				☐
				☐
				☐
				☐
				☐
				☐
				☐
				☐
				☐
				☐
				☐
				☐
				☐
Total Amount For This Month				

Expense Tracker - May 2020

Date	Description	Amount	Payment Mode	Paid
				☐
				☐
				☐
				☐
				☐
				☐
				☐
				☐
				☐
				☐
				☐
				☐
				☐
				☐
				☐
				☐
				☐
				☐
				☐
				☐
				☐
				☐
Total Amount For This Month				

Expense Tracker - June 2020

Date	Description	Amount	Payment Mode	Paid
				☐
				☐
				☐
				☐
				☐
				☐
				☐
				☐
				☐
				☐
				☐
				☐
				☐
				☐
				☐
				☐
				☐
				☐
				☐
				☐
				☐
Total Amount For This Month				

Expense Tracker - July 2020

Date	Description	Amount	Payment Mode	Paid
				☐
				☐
				☐
				☐
				☐
				☐
				☐
				☐
				☐
				☐
				☐
				☐
				☐
				☐
				☐
				☐
				☐
				☐
				☐
				☐
				☐
Total Amount For This Month				

Expense Tracker - Aug 2020

Date	Description	Amount	Payment Mode	Paid
				☐
				☐
				☐
				☐
				☐
				☐
				☐
				☐
				☐
				☐
				☐
				☐
				☐
				☐
				☐
				☐
				☐
				☐
				☐
				☐
Total Amount For This Month				

Expense Tracker - Sept 2020

Date	Description	Amount	Payment Mode	Paid
				☐
				☐
				☐
				☐
				☐
				☐
				☐
				☐
				☐
				☐
				☐
				☐
				☐
				☐
				☐
				☐
				☐
				☐
				☐
				☐
				☐
Total Amount For This Month				

Expense Tracker - Oct 2020

Date	Description	Amount	Payment Mode	Paid
				☐
				☐
				☐
				☐
				☐
				☐
				☐
				☐
				☐
				☐
				☐
				☐
				☐
				☐
				☐
				☐
				☐
				☐
				☐
				☐
				☐
Total Amount For This Month				

Expense Tracker - Nov 2020

Date	Description	Amount	Payment Mode	Paid
				☐
				☐
				☐
				☐
				☐
				☐
				☐
				☐
				☐
				☐
				☐
				☐
				☐
				☐
				☐
				☐
				☐
				☐
				☐
				☐
				☐
Total Amount For This Month				

Expense Tracker - Dec 2020

Date	Description	Amount	Payment Mode	Paid
				☐
				☐
				☐
				☐
				☐
				☐
				☐
				☐
				☐
				☐
				☐
				☐
				☐
				☐
				☐
				☐
				☐
				☐
				☐
				☐
				☐
				☐
Total Amount For This Month				

Notes

Notes

Notes

"All the money I need is flowing to me now"

Notes

"You will attract
everything that
you require."

Notes

Made in the USA
Middletown, DE
08 January 2020